CHRONICLE BOOKS
SAN FRANCISCO

how engineers see

Library of Congress Cataloging-in-Publication Data available.
ISBN-10: 0-8118-6054-X
ISBN-13: 978-0-8118-6054-3

Designed by IDEO.

Manufactured in China.

Distributed in Canada by Raincoast Books
9050 Shaughnessy Street
Vancouver, British Columbia V6P 6E5

10 9 8 7 6 5 4 3 2 1

Chronicle Books LLC
680 Second Street
San Francisco, California 94107
www.chroniclebooks.com

Contents

Foreword

Perception is the process of sensing and becoming aware. And everyone perceives the world differently. Engineers, as one group of observers, are probably overlooked and misunderstood, and certainly stereotyped as guys (yes, guys) with white shirts and pocket protectors and calculators, who are obsessed with precision. But what draws someone into the field of engineering is, in large part, simply a drive to make the world a better place, whether by focusing on the improvement of one component of a system or by trying to improve a whole system.

We engineers are inherently curious. We are natural problem solvers who are always seeking ways to perfect the things that we can influence, and we are usually not so interested in seeking the limelight. Certainly, some engineers end up as politicians or CEOs, but relatively few do compared to the number whose destiny it is to toil away in a cubicle, attempting to squeeze a little more efficiency out of a machine. The contributions of engineers are, in fact, usually invisible—and often intentionally so. It is our job to deliver an experience or an end result as seamlessly as possible.

Civil engineer Henry Petroski is well-known for writing about engineering and failure in his 1985 book *To Engineer is Human*. This book, in contrast to Petroski's work, is not about analyzing large and obvious failures; it's about noticing the subtlest engineering cues. It's about expanding the definition of a good engineer and creating empathy for a crucial but very quiet

profession. It's about learning from our mistakes and getting people to think about judicious use of materials and precious resources. It's about seeing how complex and unpredictable the world really is, and recognizing that one can never really foresee what will happen to an object once it is "released" into the environment. It offers just one of many ways of looking at the world.

In recent times, public perception of engineering has been defined more by its notable failures than by its grand successes. When those successes occur, there is no shortage of people who deserve to share the credit. I believe that the golden age of the heroic, visionary engineer has largely passed. It's hard to think of who today's Alexander Graham Bell, Thomas Edison, or Wright brothers might be. Now, that kind of honor belongs to whole teams of professionals who contribute in multiple ways to efforts too large for any individual to accomplish alone. This book is not about the grandiose world of technical achievement. On the contrary, it contains snapshots of everything you might normally pass by without noticing. It celebrates the littlest things, like rain on a car window or a loose screw, because, despite their seeming insignificance, there is still a story worth pondering—and sometimes even a lesson to be learned—behind each one.

As an engineer, I find myself constantly looking around at the world and marveling at the variety and the complexity of the things we have put here. This book is a small collection of such

observations that I and others (mostly engineers) have made as we exercised our natural curiosity. As the collection has grown in size, themes have emerged, and with those themes, some general lessons. This is a chance to step inside the skin of an engineer and to wander the world with an engineer's eyes. This is not to say that any of the observations here are the exclusive domain of engineers, but perhaps they represent one particular view of the world, held by those who share curiosity, passion for technology, and a little training in the art and science of making things work.

This book is not as much about answers as it is about questions. It is not intended to be a guide to the built world around us, but a spur to encourage us all simply to be more inquisitive. In fact, some of the explanations here are speculation (and even perhaps erroneous), but hopefully still useful in demonstrating the range of possibilities. The observations are grouped into two sections. The first half of the book shows the struggle that goes into making ubiquitous objects do their jobs and the triumph that engineers experience when the objects succeed, and it tries to reveal some of the thought processes behind their work. It provides insight into a very common question: Why did they do it that way? Only in this case, the "they" refers to engineers like me.

In the second half of the book, stories unfold. In these stories, pieces of engineering and design are deployed in the world to carry out their useful functions, beyond the protective reach of

**the littlest things,
like rain on a car window
or a loose screw...**

the people who created them. Under these conditions, objects are abused, misused, worn out, and corroded, but they are also smoothed, rounded, and worn in sometimes pleasing ways. Despite the attempts of designers and engineers to anticipate every twist of an object's fate, the unforeseen usually happens. And sometimes the engineering just wasn't that good in the first place. Looking back at the life of a building, for example, allows us to see the weakest part of the structure—and to gain information that will be extremely useful when the next one is constructed.

Perhaps we discover a point of failure that is completely counterintuitive, as when corrosion aggressively attacks the most protected part of a steel beam. And we can also see success, when things do go as planned and the end product proves to be a match for everything that is thrown at it. Regardless of whether we find inspiration or not, we owe it to ourselves and those around us to become better observers. Our environment is brimming over with information that can help us with our basic ability to navigate a course. The better we are able to refine our actions and our thoughts based on seeing what has gone before, the fewer mistakes we will make and the world will be a better place for it.

At the most basic level, an engineer is trained to simply make things work. That alone can be enough to keep huge numbers of engineers productively engaged. But as products are gradually refined through iterations over time, the engineer's task becomes one of making things work *better*. This is a different kind of challenge, and it requires optimization of resources and a focus on safety, cost, and efficiency. To get the most out of limited resources, engineers must navigate through a host of conflicting constraints, with choices at every turn. Their decisions are ideally informed by data to ensure the repeatability and reliability of their results, but the answers are only as good as the questions, and so each design follows a path mapped as much by intuition as it is by calculation. While it is the designer's job to set out a vision for an end result, it falls to the engineer to sort through the options for making that vision reality: options like strength versus weight, durability versus cost, and form versus function. A good engineer manages to balance these factors while maintaining the integrity of the original vision. The more constraints, the more difficult—but also the more rigorous and precisely matched—the solution will be. You can see this balance played out in the world around you.

Part 1 : Creation

The making of things

Craftsmanship

The human hand in a machine-built world

Every manufactured object, by definition, has been touched by the hand of a person in one way or another. Even the simplest of things, such as a paper clip, is the end result of significant creativity and a chain of decisions, as well as the persistence that was required to determine its shape, plan its craftsmanship, and then construct the equipment needed to make the object by the millions. Despite the achievements of the Industrial Revolution, it's still possible to discern the original thought, and then the manual execution of that thought, in the things around you. While modern manufacturing processes relentlessly pursue uniformity, it is the little imperfections that remind us of the mundane but necessary roles that people play in making things. If we could eliminate such imperfections altogether, would we lose a vital connection and start to believe that people are no longer necessary in the process of making things? If machines were to produce everything, would we have to accept the "craft of the machine" as some new benchmark of aesthetic warmth and quirkiness? Can a mold-mark take the place of a fingerprint? If we could eliminate all possibility of human error (which is impossible), we might also eliminate any sense of responsibility for the results of our actions. Pride in the skill of the individual would be gone, and in its place would remain only admiration for the infallibility of the machine.

ST
T
HERE

Unseen

What we can’t or won’t see

Façades are found in many places, most traditionally on buildings, where they are erected to project an image that is more stylized or refined than the structures behind them. Trickery, in a way. The concept of the façade is prevalent in the objects and even the services we use every day. Turn on the water and a hidden, often ugly, system provides (one hopes) a clean, consistent flow. Over time, we have become used to reaping the benefits of such infrastructure without giving the infrastructure itself a second thought. These systems are often huge and complex entities, sometimes many times larger than the façades that hide them. While it's easy to focus on the tangible face of a product or service, it's interesting to seek out, or at least to imagine, what is going on behind the curtain. We have an opportunity to do just that when an object or service is demolished, corroded, or otherwise highlighted—sometimes by something as simple as a flow of air—and its internal workings are thus brought to light. Perhaps more difficult to see are those elements of machines that are in plain view but intentionally hidden or camouflaged by use of lights, paint, or scale. Sometimes you just have to take a closer look.

Gaining an appreciation for the underside of technology generates bigger questions. As resources become ever more scarce, can we afford the luxury of hiding things from view? On the other hand, is it inherently good to know how things work? Should we move toward exposing inner workings in an attempt to raise awareness and promote judicious usage? Or is it unsettling to see what it takes to provide us with the necessities we are used to?

Unseen | 23

TRAFFIC

PLEASANT PRAIRIE WI
PLEASANT PRAIRIE WI

DU PONT
Tyvek
HomeWrap
Call 1-800-44TYVEK WWW.TYVEK.COM
DU PONT The miracles of science

Unseen | 27

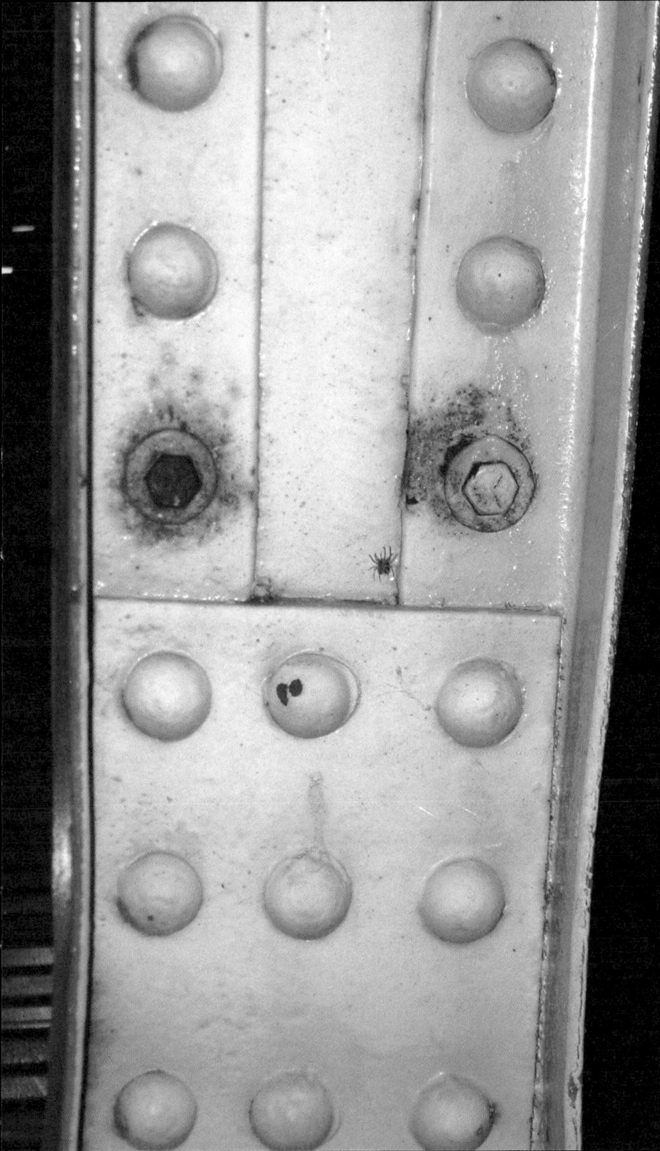

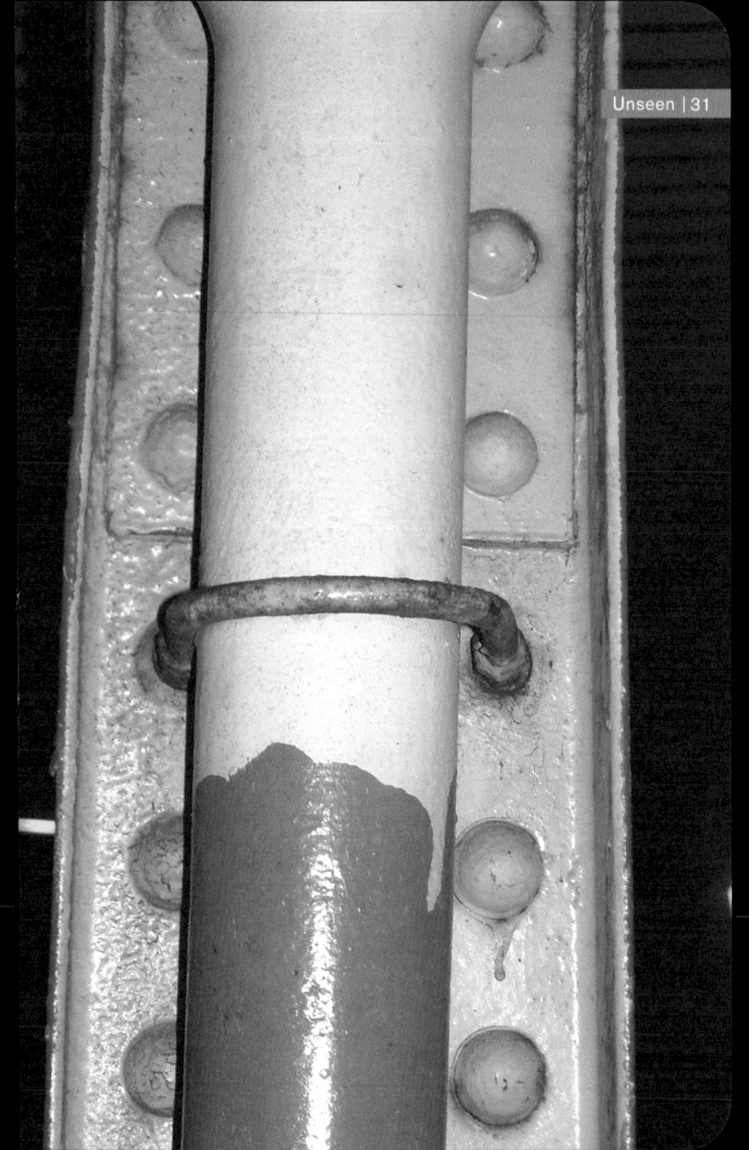

FARROW
FARROW

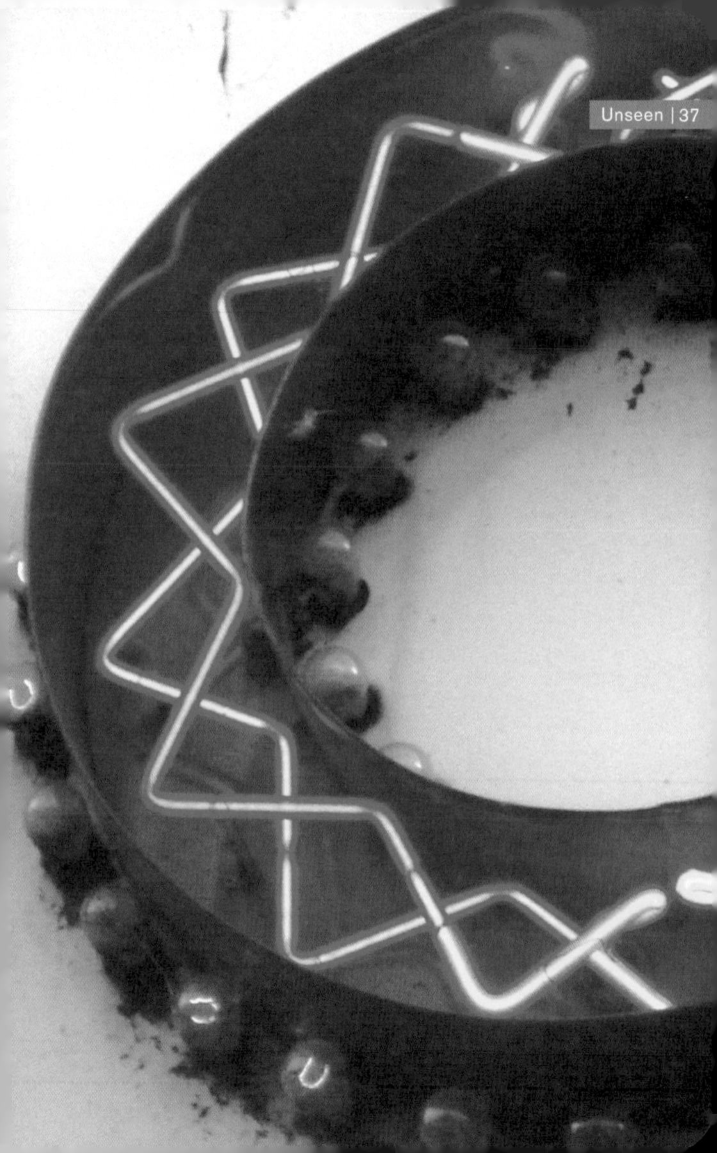
Unseen | 37

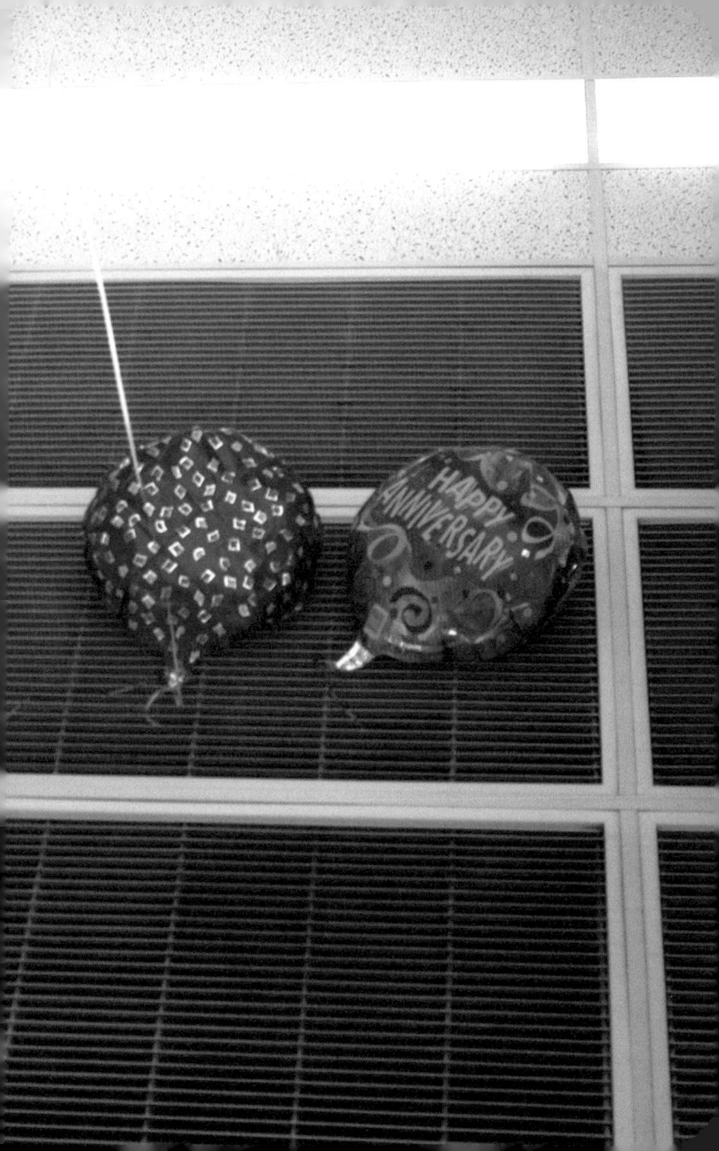
HAPPY
ANNIVERSARY

Unseen | 39

Illusions

Anachronistic idiosyncrasies

An optimal solution may not be an intuitive one. When things are designed, the most intuitive path or solution is usually the one that is chosen first. Over time, improvements are made in design, materials, cost, and manufacturing. As a result, the design of an object often evolves away from that first intuitive idea. This results in interesting contrasts, especially when you can see the "before" concepts and those that came afterward, side by side. Jarring or counterintuitive changes to the object, such as replacing material or elements with stronger components, changing constraints, or refining structures, can create surprising results—for example, bridges that span greater distances, or foundations that eliminate the need for extra supports—leaving the observer to rely more on trust than on intuition for his feeling of safety.

ACROSS AMERICA!

Interfaces

How things connect

Psychologists study interactions between people, and ergonomists maximize the efficiency of people's interactions with their work, but it is the engineer's job to make objects work well with each other.

An interface is a meeting point between two things or people. The most likely of such meetings can be predicted, but there are always surprises. Some interfaces, the ones that are anticipated, get careful consideration in the design process, while others are overlooked, which leads to the need for adaptation. Many items are made with only a limited brief in mind: its immediate perceived use and its technological context in the time when it is made. It is often difficult or impossible to account for the numerous alterations to an object's use, usefulness, or environment that will be made in its lifetime.

What happens when two objects or surfaces unexpectedly come together for the first time and someone must make them fit? And why does this happen so often? Some answers may be found in the interesting trade-offs made in the design process. The more precisely an interface is crafted, the better it fits and works, but the less adaptable it becomes. For example, a bolt and a nut can be designed to fit together precisely, but only if they are of exactly the same specification. Conversely, the more generic and versatile the interface is, the poorer the job it does of creating a secure connection, leading to on-the-fly refinements that provide additional features. (Why else would duct tape exist?) Both approaches have their merits, and both may rate highly in terms of sophistication or simplicity, but because there are so many less-than-perfect interfaces, one wonders if an entire profession could be found in simply making things fit. In the worlds of engineering and fashion, there is indeed such a profession—known as a "fitter"—and its very existence proves that precision, when it comes to interfaces, is very hard to design for.

Elegance

Balance, harmony, efficiency, and simplicity

What is elegant to an engineer may not seem very beautiful to everyone else. Here *elegance* means a judicious and efficient choice of materials or processes to achieve a goal. Engineers are not renowned for their sense of aesthetics—on the contrary, they have a reputation for being obsessed by a design's functional aspects at the expense of its beauty—but let's dig a little deeper. Isn't one definition of *beauty* a pleasing balance of elements that work well together? If so, then engineers do create beauty, using attributes that are not typically recognized as beautiful to look at. They strive to make things blend and work seamlessly in terms of material use, energy efficiency, abrasion, corrosion, and cost. Elegance is often associated with simplicity, which is among engineers' more difficult accomplishments. It's relatively easy to add parts to a system to make it work, but it's difficult to take elements away while preserving the system's required functionality. Time is money, and so time and care are not often spent to reduce designs to their simplest forms. Yet when all the conflicting design variables are optimized and harmonized, the result embodies elegance, and elicits a reaction in an engineer not too different from that of an artist looking at a fine painting.

SUGINO 75

MADE
DUMP
TO
DRAINS
EAST

IRON WORKS
WASTE!
TERWAYS
SA

HAND PICKED BEANS

Function follows form

The interplay between design and engineering

Engineering is often called upon to help realize a designer's creative vision while utilizing the most efficient and economical materials and structures. But often (too often) one aspect dominates the other. A designer may envision a plan without taking adequate time to consider how it might be realized. When this happens, the design almost inevitably will be compromised when its functional aspects are resolved. This is often easy to spot, since the design embodies the outward-facing part of a product, while the functional contributions are usually less obvious. And of course this problem can work the other way, too. Many products start life as "inventions" based on some incredibly clever piece of technology, and make the way through the development process—and almost to the point of sale—before people realize that they need to be usable and attractive as well as functional. In these cases, design is layered on at the last minute. The best results occur when design and technology work hand-in-hand, when each point of view influences the other. This requires a partnership between art and science.

PROGRAMME

13

12

11

10

9

8

4 15 • 1 2
4
5
6

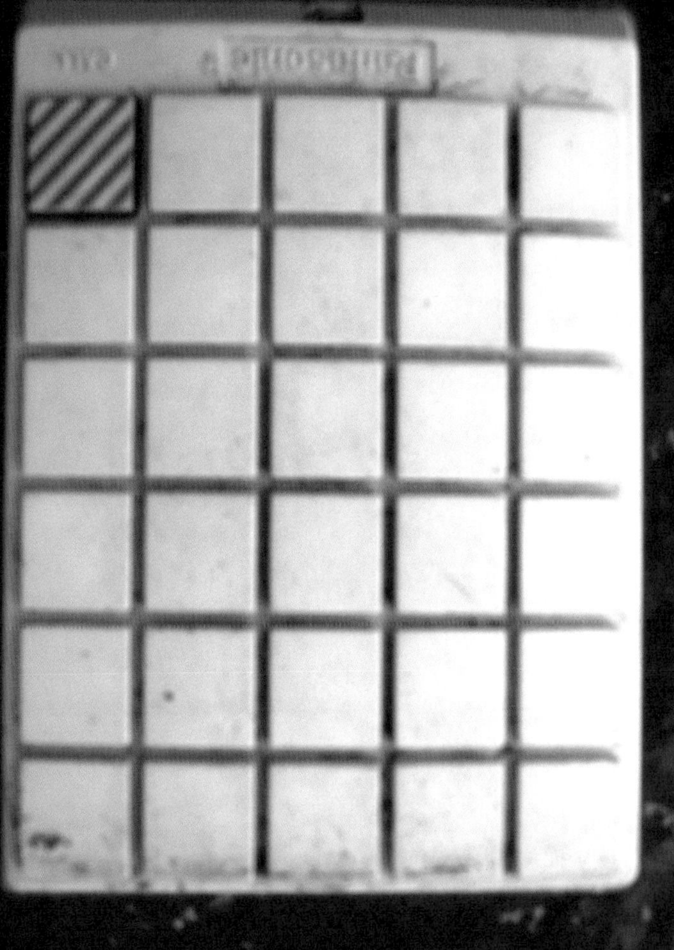

4

FIRE
EXIT

20
18
16

20
14
VATAN

Sequences

Unraveling the chain

Most constructions are made in serial processes. Step 1 is always followed by step 2, and so on. And when one process is finished, another comes along on top of it. Over time, layers build up, either literally or in a ripple effect, when the first element establishes itself and subsequent ones have to work around it. Careful observation can reveal the sequence in which things were completed, and you can imagine the thought processes driving the sequence, the decisions about whether to work around something or make a fresh start. Some of these observable choices seem logical, while others clearly are not ideal and were derived from conflicting processes or arrangements. Yet these conflicts lead to ideas for potential improvements in the way things get done; if there's a lesson here, it's "Plan ahead." Of course, planning ahead sometimes requires the ability to see into the future, but at other times, perhaps only a little more careful forethought is needed.

NO PARKING HERE TO CORNER
NO PARKING 15 MIN LOADING ZONE
STOP

Challenges

"Why can't they fix that?"

Engineers like to grapple with tough problems. Such problems arise from a project's inherent constraints, and it is the job of the engineer to analyze and juggle those constraints, ultimately arriving at the optimal solution. Sometimes the solutions are obvious and even nonprofessionals can clearly see that they work well. There are also cases where there is no ideal answer, and we are left with something that is as good as it can be, but not really good enough (and certainly not elegant). In such cases, the compromises are more obviously perceived, and they provoke comments such as "Why can't they fix that?" In still other solutions, the struggle to balance competing constraints is hidden a little deeper, or the path to the ideal answer is blocked by an annoying obstacle, such as a patent owned by someone else.

Creating new objects, machines, buildings, and all the other things that we have come to rely on is a very hands-on process. Lots of attention and care is generally given to every detail, from the original concept through manufacturing and finally to packaging, sales, and the point in time when the entity reaches its final destination and is used. Along this path, engineers and designers exert control to keep things on track, but eventually control is relinquished and ownership is passed on. Whether the engineered object is a spoon, a train car, or a brick that forms part of a wall, it is required to live its useful (and ideally, trouble-free) life until it is worn out or exhausted.

The designer and the engineer work hard to anticipate the circumstances that their product will encounter, with varying degrees of success. Some things are harder to design for than others. A warranty is a promise to a user that an object will enjoy a minimum period of trouble-free service, and yet it is also evidence of the trade-offs made in the design. It says "Given the constraints that surround this particular object, we're pretty sure that it will last for a given number of years." And its unwritten message says "If you were willing to pay more, we could have made it last longer," or perhaps "We designed this product to last just this long, so that when it breaks, you'll buy another one, and we'll be able to stay in business"—the concept of planned obsolescence.

No matter what the intention, the passage of time has an uncanny ability to reveal some weaknesses that were impossible to predict. Materials behave in unexpected ways when exposed to unforeseen conditions. We can see how Mother Nature slowly, or sometimes quickly, tears things down, and we marvel at the complex interactions that occur out there in the "real world."

Part 2 : Degradation

The destiny of the things we make

Corners

No good way around them

A corner is a change of direction in a surface or a line, and corners are inescapably problematic. Sharp internal corners create weaknesses that lead to cracks. External corners are subject to damage and can be nuisances—and yet we still have to use corners in all sorts of places. The most basic engineering classes teach students how weak sharp corners are and that eliminating them is the best strategy. Using a sharp corner is, unfortunately, usually cheaper and easier than going to the extra work of rounding it off, despite the known superiority of rounded corners. Aesthetic concerns also come into play here: a crisp edge may look better than a soft one in a given design, or it may provide more secure alignment for a feature. But in the long run, numerous failures, small and large, can be traced to sharp corners breaking, and numerous injuries can be attributed to impacts with those same sharp corners.

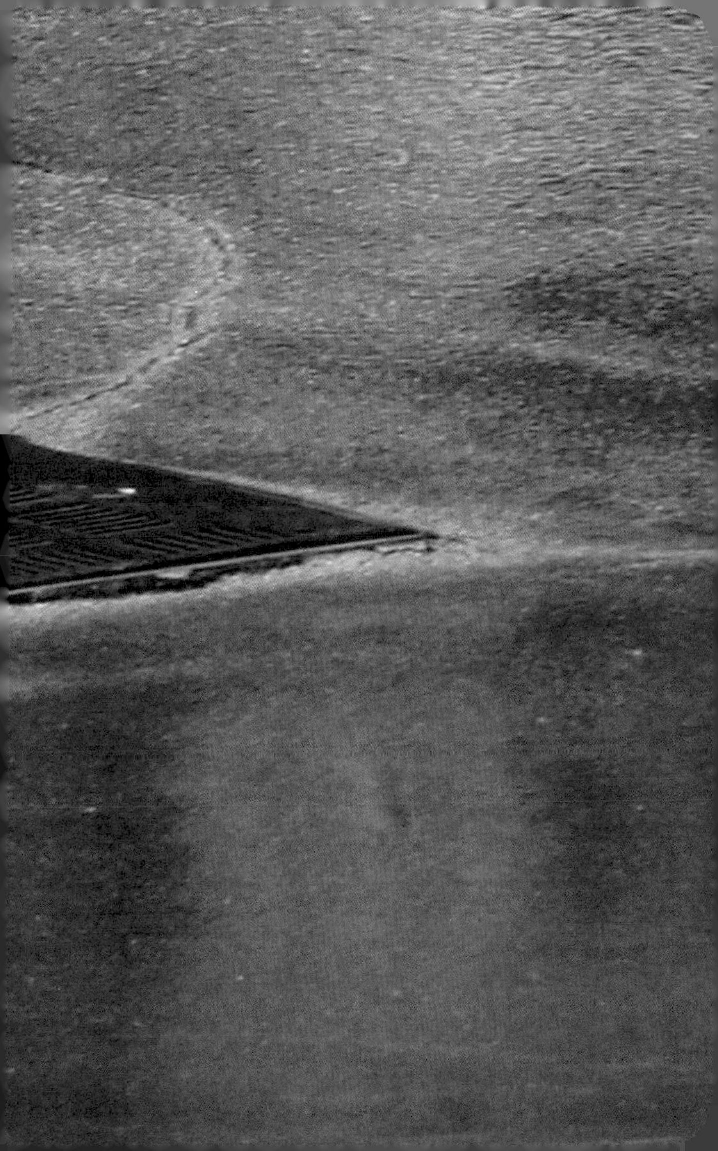

Ugliness

When beauty is not a priority

Shortcuts or focusing on the wrong details can lead to ugly end results. Or maybe making an object look good is just too expensive. People operate with different priorities; perhaps speed, cost, service, convention, or training becomes the overriding factor in their choices. And people are usually taught to operate the same way each time—quality control is stressed, consistency is valued, change is discouraged. Making something look nice can take extra time, as can finishing off a detail with care or trimming away excess material. And time is money. Just as what is elegant to an engineer may be invisible to others, what is ugly to everyone else may be invisible to an engineer.

UNITED STATES
POSTAL SERVICE

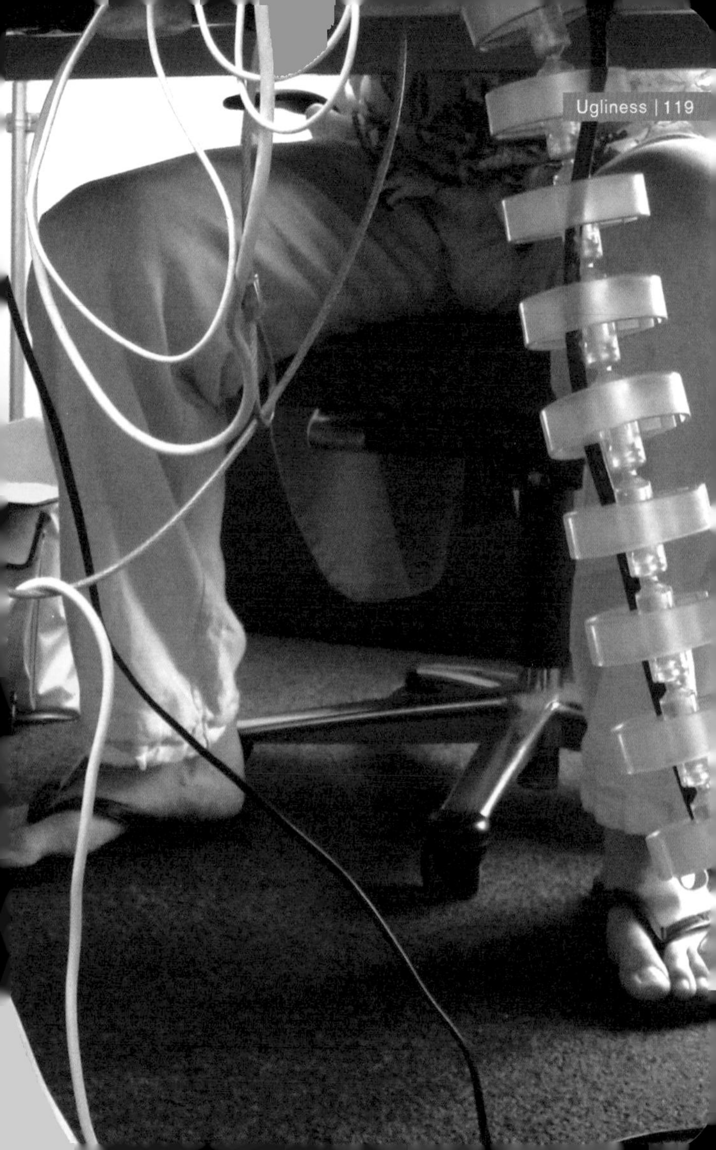

AC-630 TC

Materials

What we have to work with

Everything is made of one material or another, and increasingly our materials have lots of interesting properties that manifest themselves in unique ways. As these materials are exposed to external influences such as wear and tear, corrosion, general use, and the elements, they start to change. These changes might be gradual or sudden and permanent, and they might be reversible under different conditions. The nature of these changes gives us clues to the age, usage, and durability of the materials themselves and provides us a glimpse of an object's history.

COAT HOOK

11 FT 7 IN

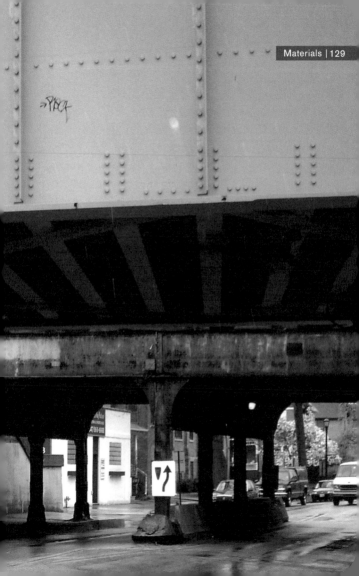

Repetition

Persistence yields results

A process's singular actions often appear to have no effect, but when they are repeated over and over again, there is a noticeable cumulative result. Such traces can reveal trends or interesting phenomena that otherwise might not be visible. Repetitions can happen slowly, revealing their secrets over years or centuries, or they can happen very quickly, as in the form of vibration. An impact or force that repeats itself as often as multiple times a second accelerates events and effects, such as the loosening of a screw, that otherwise would happen much more slowly.

U.S. COINS
LOUD

COIN RELEASE

1
ABC 2
3
GHI 4
JKL 5
MNO 6
PRS 7
TUV 8
WXY 9

PrtSc
SysRq
NumLk
ScrLk
Pause
Break
Insert
Home
PgUp
F9
F10
F11
F12
Delete
End
PgDn
Backspace
U
I
O
P
J
K
L
Enter
N
M
Shift
Alt
Ctrl

Consequences

Small actions can have big effects

Hasty solutions, poor improvisations, and lack of the right tools or equipment can result in experiences that range from disappointing or off-putting to downright jarring. Effects like these may be immediately obvious or may reveal themselves over time. Such failures can be attributed to a lack of foresight about how things will be used. Some mistakes may even be the result of deliberate actions (such as a conscious choice to pick a less durable material for cost reasons), but many are a consequence of lack of attention to details. Getting out into the world and looking at how things really get used and abused adds new understanding of what's important. Design and engineering decisions that may seem unimportant can take on greater significance when circumstances amplify an object's weaknesses or shortcomings.

Nature

Forces we can't control

Machines, technology, products, and buildings all exist in the natural world, where nature intervenes or interacts with them. Although nature is increasingly under pressure from the advance of the human race, there is a counterpoint to that pressure. When an object is placed in an environment, degradation of some form immediately sets in, and gradually, inevitably, the unnatural world is engulfed by the natural one. Living organisms and the atmosphere combine to clog, fade, split, warp, rust, tangle, and cover whatever we put out there. It is fascinating, and perhaps even encouraging, to see that the best we have to offer is never quite good enough to prevent nature from returning everything to its most stable form. Warranties and guarantees are designed to protect consumers should this happen too early in the lifetime of a product. (And why do we speak of a "lifetime" for inanimate objects, as if they were in fact alive?) The manufacturing processes we have evolved change materials and elevate them to a more useful form, but this form is still only transient, whether it lasts seconds or millennia.

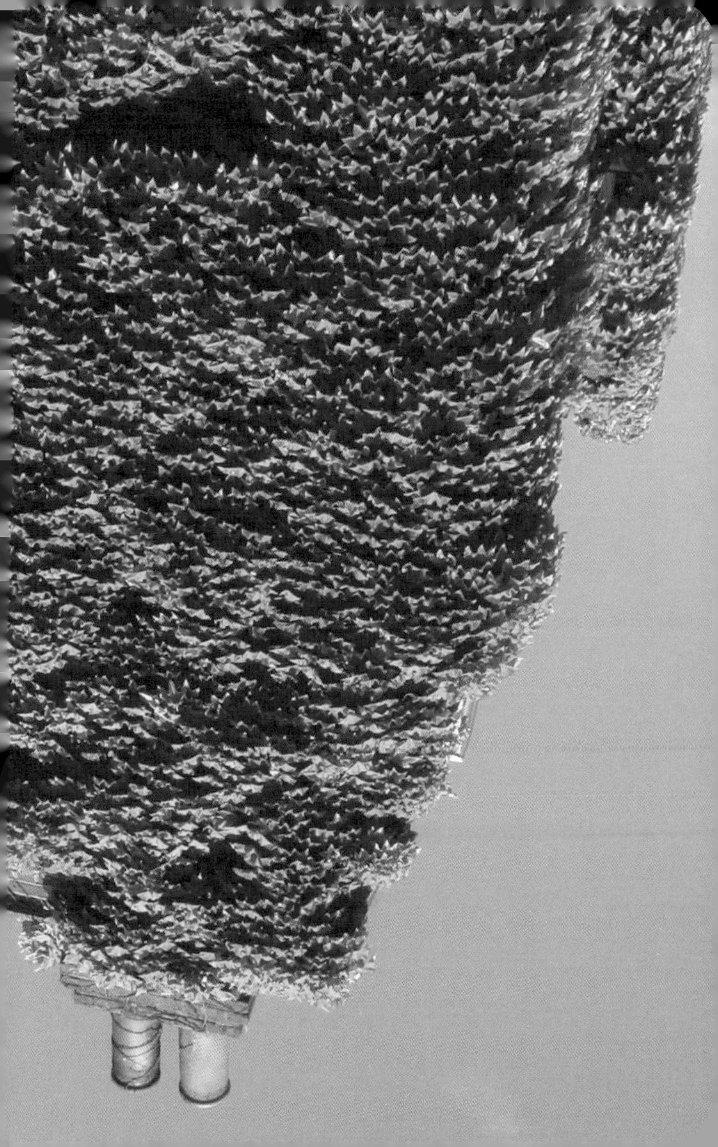

Time

Lessons from the future

Structures are subject to the constant pull of gravity, the wear of use, and the battering of the elements, and sooner or later they will succumb to these forces. With the benefit of hindsight, we sometimes can look at a failure and see where its design could have been stronger. By examining objects after significant time has elapsed (or when time has been accelerated for testing purposes), we can compare their end to their beginning and learn valuable lessons that enable our future designs to be stronger and more durable, or simply to age better. For designers and engineers, looking at old things is a way to glimpse the future—to see what to do and what not to do—which is a rare opportunity indeed.

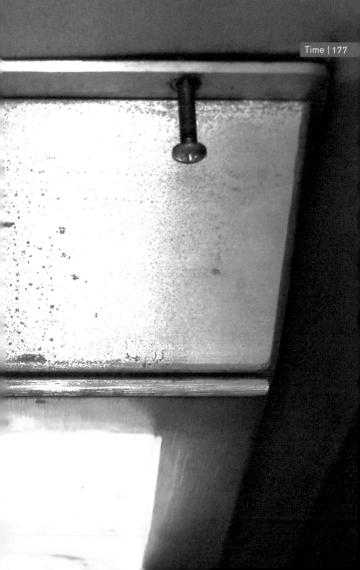

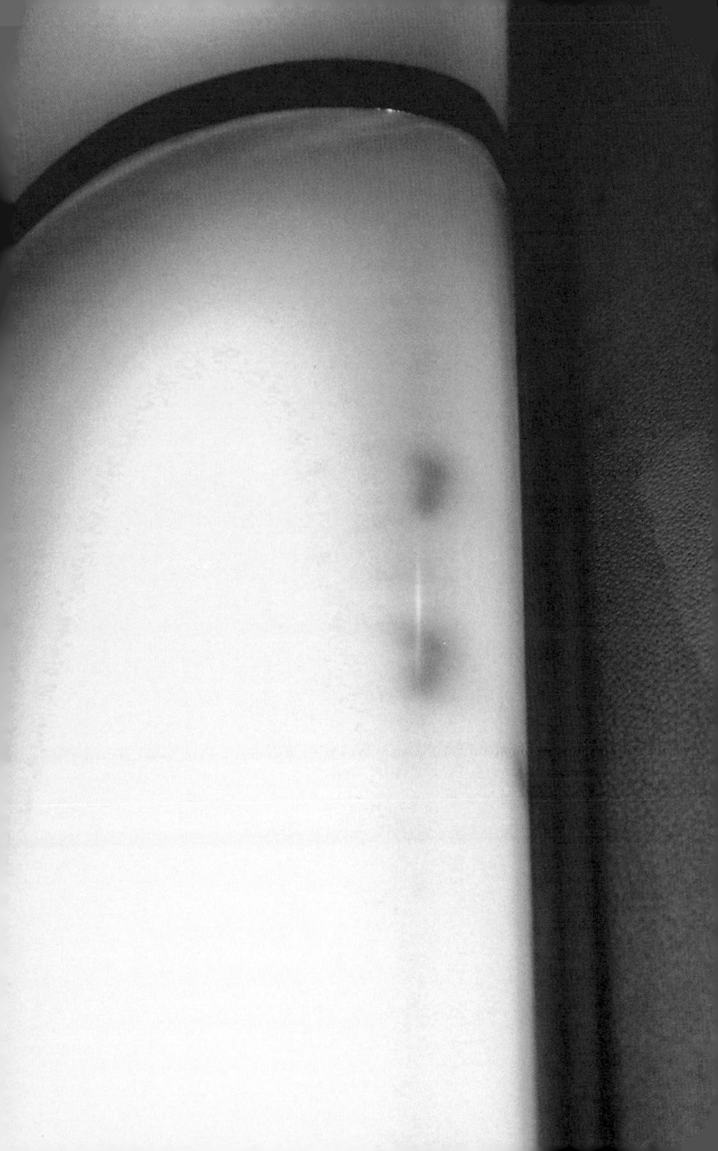

Time | 181

FOREVER

Index

Looking back

Craftsmanship | 12-19

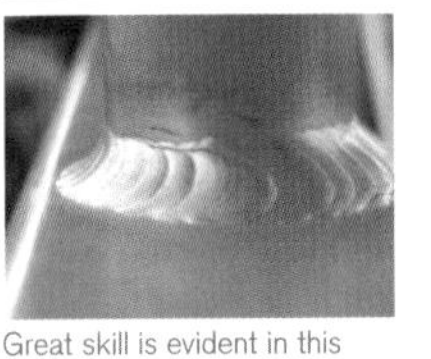

Great skill is evident in this hand-welded joint.

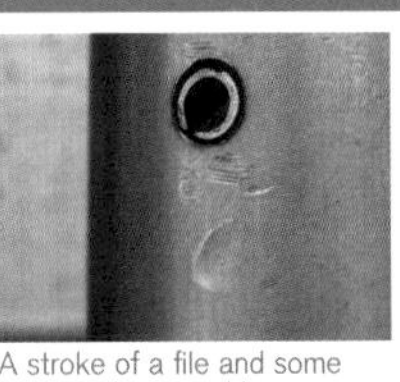

A stroke of a file and some careless hammer blows are frozen in time.

Hand lettering can quickly and cheaply be customized to fit any space.

Unseen | 20-39

Haphazard fasteners reveal the underbelly of 1980s automotive manufacturing.

Fire hydrant connected to an underground network of pipes, ready for action.

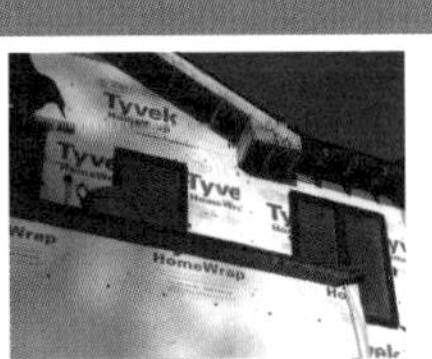

A strange, branded layer, only visible during construction.

A forgotten phone closet also houses newer, more compact technology.

Connections and sockets are in plain view, but masked by the bright lights.

To the right, the fairground experience; to the left, the infrastructure.

Illusions | 40-49

Huge steel beams allow this bridge to hang over the road without visible supports.

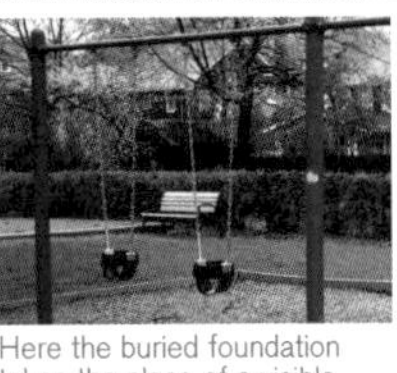

Here the buried foundation takes the place of a visible frame.

A-frame structure that is intuitively stable.

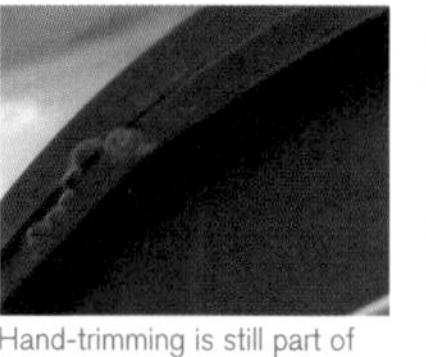
Hand-trimming is still part of this automated manufacturing process.

The laborious and pains-taking wiring inside an ancient telephone closet.

Reinforcing steel is the last thing left when a building is torn down.

Windows that no longer bring natural light or ventilation.

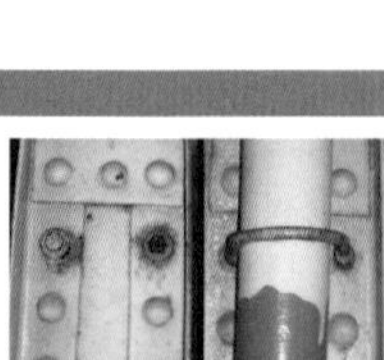
The telltale rusted bolts signal something hidden, and on the back is the pipe bracket.

From a distance, the letter O, but up close, an intricate pattern.

Drifting helium balloons reveal the otherwise invisible airflow.

A rare look at the inside of an aerosol can during decomposition.

Not sagging under its own load, this catenary curve is the most efficient shape.

Rust damage occurs, perhaps where the sun never reaches to dry things out?

These boats appear to be flying out of their usual context.

Interfaces | 50-59

The connection between rail and bracket seems improvised.

The feet on this bench could easily be designed to be placed on soft ground.

An interface that has been carefully developed and refined.

A connection designed to be permanent but since converted to be portable.

Square peg in a round hole.

Elegance | 60-71

A classic design that has remained unchanged for more than 100 years.

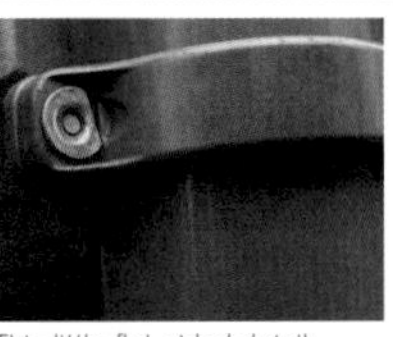

This little flat-sided detail eliminates the need for an extra tool.

Small cover plate saves the time otherwise needed to align these two fridges.

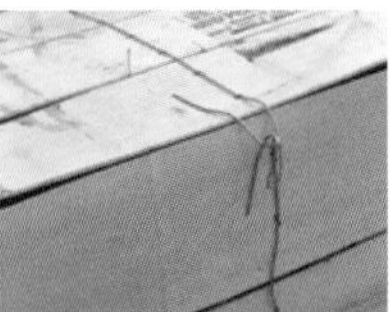

The wire provides strength, hinge and lock for this box.

An interface hastily assembled using at-hand materials.

Spur-of-the-moment modification just to make it work.

Lamppost base flange that can only fit a perfectly flat sidewalk.

The shape prevents the cover from falling in the hole, while its weight keeps it in place.

This wire knitting results in a strong and lightweight fence.

Better to let this bridge pivot than to try to hold it down.

Function follows form | 72-81

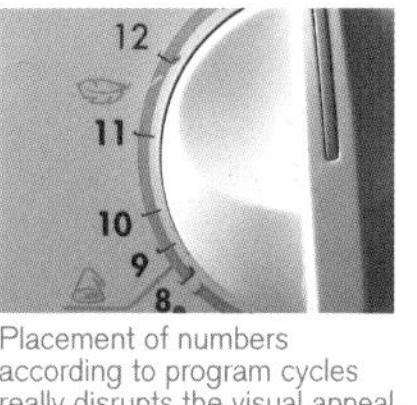

Placement of numbers according to program cycles really disrupts the visual appeal.

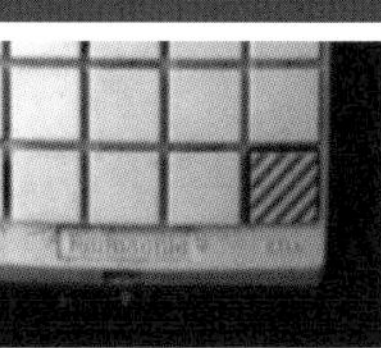

This expandable doorbell system is clearly struggling to make aesthetic sense.

The geometric layout of this door is not able to maintain its shape without help.

Sequences | 82-91

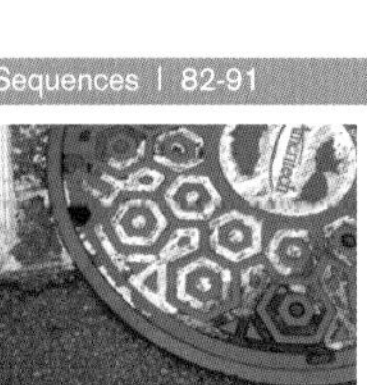

The cover has been painted several times but still is not aligned.

The sign is placed in a ready-made slot and then twisted into the right orientation.

Unanticipated dish antennae reveal how many satellites orbit overhead.

Challenges | 92-101

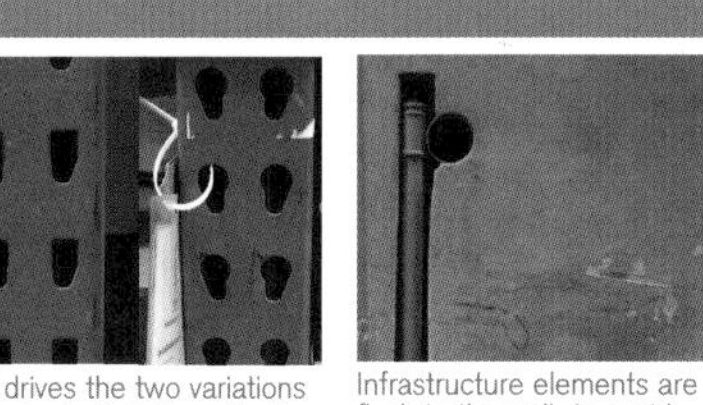

The wet windshield triangle symbolizes compromise between economics and engineering.

What drives the two variations of this design? Is it aesthetics, strength, or patents?

Infrastructure elements are flush to the wall, to avoid getting hit by traffic.

Corners | 104-111

This grating has street cracks on all four of its sharp, square corners.

Broken ground-pin holes are a problem that was designed-in and replicated.

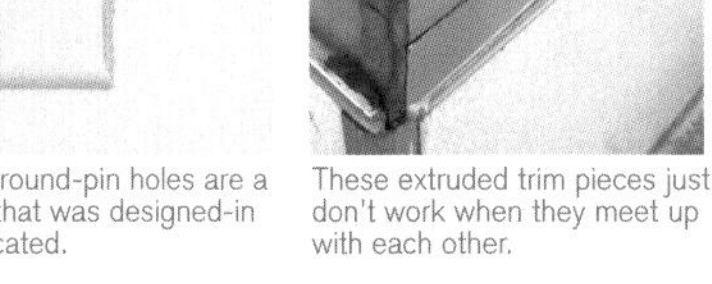

These extruded trim pieces just don't work when they meet up with each other.

What choice was there here, but for the door to fit the building?

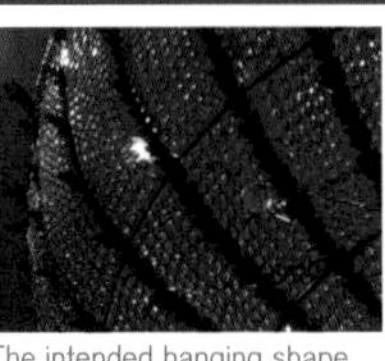
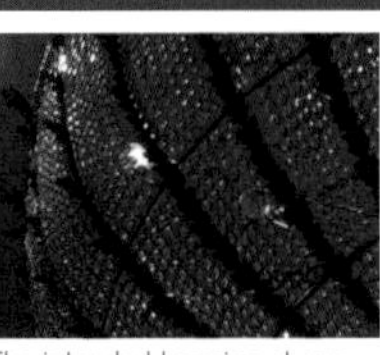

The intended hanging shape of this chandelier needs support by a rigid structure.

This dredger is all function and not much form.

The addition of air circulation was not factored in to the original design of this building.

The best solution to work around the water connections was to split the sign.

The unusual shape of these slats allows an even gap, even as they hug the metal arms.

This egg carton is afforded more stability by adding corners to four of its feet.

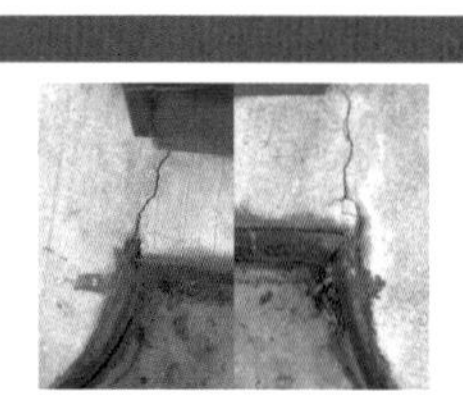

Another example of a designed-in flaw.

Ugliness | 112-121

A hauntingly quiet partially built community—someone's huge mistake.

Redundant wiring and unsightly terminations. Out of sight, out of mind?

Surely there's a more elegant way to replace the old utility pole than this?

Materials | 122-133

This metal duct has recorded the detail of each scuff and blow.

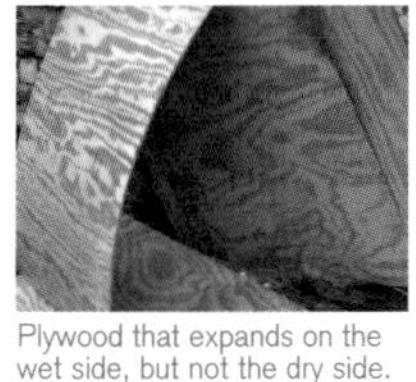
Plywood that expands on the wet side, but not the dry side.

Wear patterns on the most exposed surfaces.

Steel and glass don't mix well.

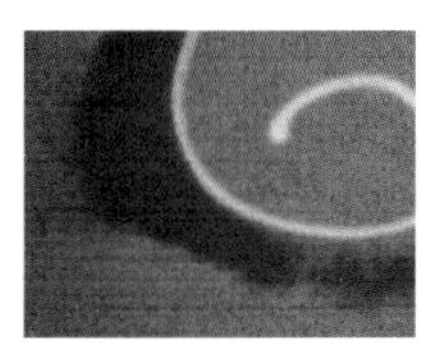
The sidewalk moisture here flowed around the water-repellent sad face.

Repetition | 134-145

Years of two-way traffic have nudged these bricks into an S-shaped pattern.

Years of use reveal usage patterns and technique.

The worn-through floor finish shows the entry/exit path of many feet.

Is the United States Postal Service using up old cans of paint?

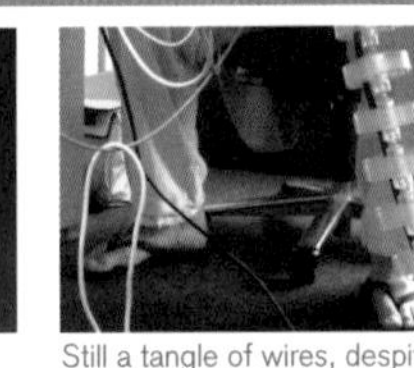
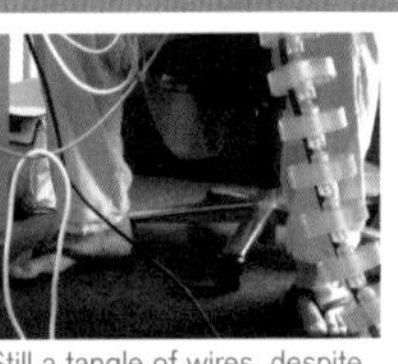

Still a tangle of wires, despite the presence of a cord-management system.

Why do the pipes have to enter from the top? And why so many pipes?

Is this an older section of drainpipe spliced into a newer installation?

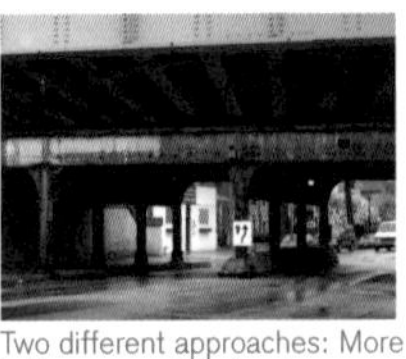

Two different approaches: More bridge supports using less material and vice versa.

The rust just keeps leaking through the paint.

Is this door handle too close to the door?

Damage caused by dripping water is unmistakable.

The vibrations from this passing train dislodge the hanging droplets on the wires.

Consequences | 146-159

Gravity works against this keyhole cover—could the pivot have been in a different place?

Is this sidewalk too narrow? Too much salt in winter?

Small errors are sometimes amplified and become more obvious.

Insulating tape is needed to keep this locomotive power connector in place.

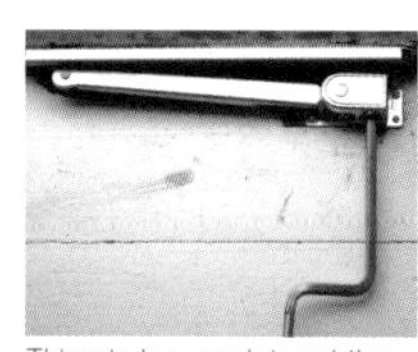

This window crank is not the best choice for this location.

The oil attracts dirt, which increases wear.

Nature | 160-173

Ivy taking over.

The deeper recesses (shadows) allow more water to seep in behind the brick faces.

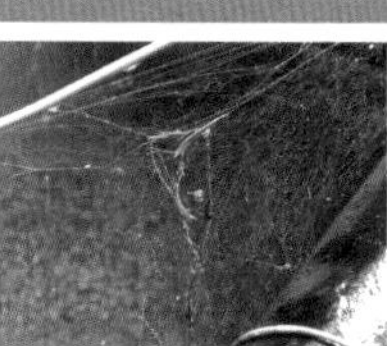

Man-made and natural fibers work together.

A chance wind cast this leaf in cement.

Fungus growing from a stone wall. Is there organic material under the paint?

A last-minute choice of washers for this flagpole detracts from the clean appearance.

A couple of missing bolts on this flange could lead to premature failure.

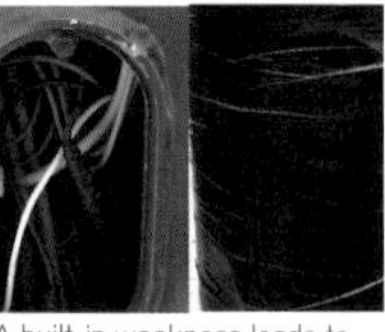

A built-in weakness leads to multiple failures.

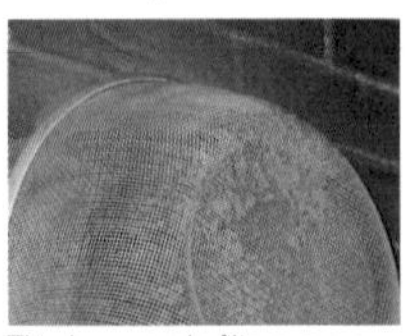

This homemade filter serves to attract attention to the dust that is being expelled.

Trees will be trees.

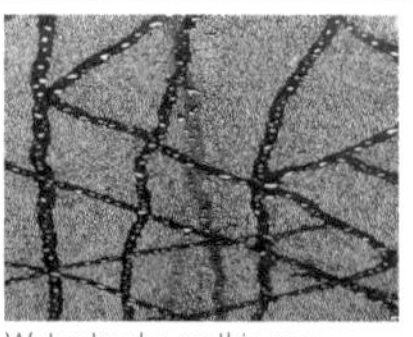

Water tracks on this rear windshield describe a tortuous journey.

Just a small foothold, but it's a beginning.

Many train trips have caused this screw to slowly loosen.

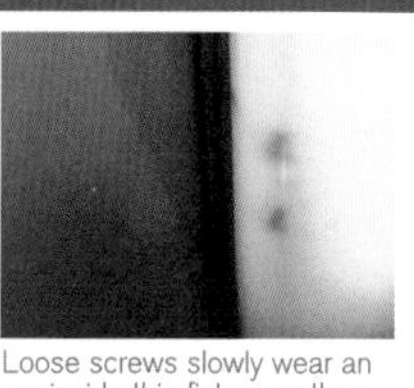

Loose screws slowly wear an arc inside this fixture as they rock back and forth.

Years of use have worn this seat down to the stuffing.

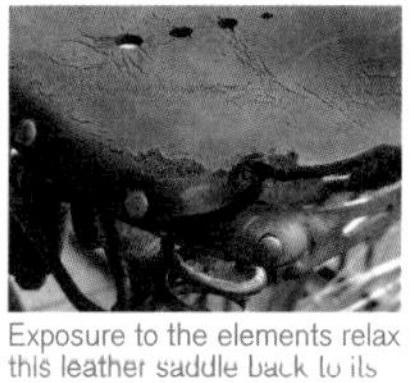

Exposure to the elements relax this leather saddle back to its original shape.

The rust here is slowed only a little by the ironic "forever" sticker.

Two generations of barbed wire, side by side. Is one there to replace the other?

This roof eventually sagged, revealing the weakest part of the building.

Repair and maintenance of this roof have gradually built up its diverse character.

The collapse of this wall was inevitable, but hard to predict.

Afterword

Inspiration for collecting the images that led to this book came from a conversation with Jane Fulton Suri, director of human factors at IDEO. We walked to a coffee shop, and I pointed out to her the things that I saw through the eyes of an engineer. She spurred me on to do something with these observations, and so using her book, *Thoughtless Acts?*, as a model, I began to pursue a collection of causally taken, but highly informative, photos of these subtle observations. As the word got out, pictures started to come in from across IDEO, and the collection grew. Initially, the collection was shared in the form of cards that were made to simulate Polaroid photographs, with the intention of portraying the snapshot nature of the photography. A positive reception led to the concept for this book, and here it is.

Acknowledgments

Thanks go to lots of people, especially Jane Fulton Suri for the initial inspiration, and Whitney Mortimer and Caroline Herter for thinking big. Alan Rapp at Chronicle Books, for editing a non-author's work. My wife Michelle and my children Graciela, Juliana, and Philip for being patient every time I stopped the car to take a picture and for help and encouragement in seeing it through. Credit for the pictures goes to many at IDEO; in particular, Erin Koch, Ken Ritsher, Mark Fisher, Steve Schwall, John Grimley, Maura Shea, Adam Reineck, Graham Findlay, Roby Stancel, Mark Zeh, Emily Ma, Tom Cohlmia, Nathan Parkhill, Jenny Comiskey, Dave Privitera, and Yona Belfort.

About the Author

Andrew Burroughs has always been an engineer at heart, from his early exploits into undersea exploration with a cardboard aqualung (at age 6) and home-built wood-fired pottery kilns to his determined efforts to fix broken appliances and toys rather than throw them out. Like many engineers, a fascination with how things work and a passion for just making stuff kept him out of trouble. Not content with a purely technical track however, he sought education at both London's Imperial College of Science and Technology and the Royal College of Art. Upon graduation, a taste for adventure took him around the world but it was love that ultimately brought him to the United States, where he lives in Wisconsin with his wife Michelle and three children. Twenty years as a consulting design engineer (fifteen of those with IDEO) have seen him involved in a wide variety of projects. In his professional career he has specialized in the healthcare field, designing award-winning drug delivery devices, surgical tools, and a kidney transporter, among other things. Since 2004, Andrew has led IDEO's Chicago office, but still finds time to seek out the unusual and the overlooked.

how engineers see.